SO-ARV-712

Water, Water Everywhere

A Book about the Water Cycle

By Melvin and Gilda Berger
Illustrated by Bobbi Tull

CHELSEA HOUSE PUBLISHERS • PHILADELPHIA, PA

The authors, artist, and publisher wish to thank the following for their invaluable advice and instruction for this book:

Jane Hyman, B.S., M. Ed. (Reading), M. Ed. (Special Needs), C.A.E.S (Curriculum and Supervisory Development)

Rose Feinberg, B.S., M. Ed. (Elementary Education), Ed. D. (Reading and Language Arts)

R.L. 2.1 Spache

First hardback edition published by Chelsea House Publishers in 1999.

Printed and bound in Mexico.

Library of Congress Cataloging-in-Publication Data

Berger, Melvin.
    Water, water everywhere / Melvin & Gilda Berger; illustrated by Bobbi Tull.p. cm.—(Discovery readers) Originally published: Nashville, Tenn.: Ideals Children's Books, c1995. Includes index.
    Summary: Explains the cycle of evaporation, condensation, and precipitation that provides fresh water to the earth and describes how this supply is brought to people's homes.
    ISBN 0-7910-5069-6 (hc)
    1. Hydrologic cycle—Juvenile literature. [1. Hydrologic cycle.
2. Water supply.] I. Berger, Gilda. II. Tull, Bobbi, ill. III.
Title. IV. Series.
GB848.B46 1998
551.48—dc21   98-26955
                                                                CIP
                                                                AC

Turn on the faucet.
Out flows the water.
Drop after drop after drop.
You fill a glass.
And take a drink.
The water tastes good.

Here's a surprise.
The water is not new!
It has been on the earth forever.
If we are careful, there will always be
    plenty of fresh, clean water.

But where does the water come from?

Each drop of water is on a never-ending
   journey.
This journey is called the **water cycle**.
The water cycle gives us the water that
   we need.

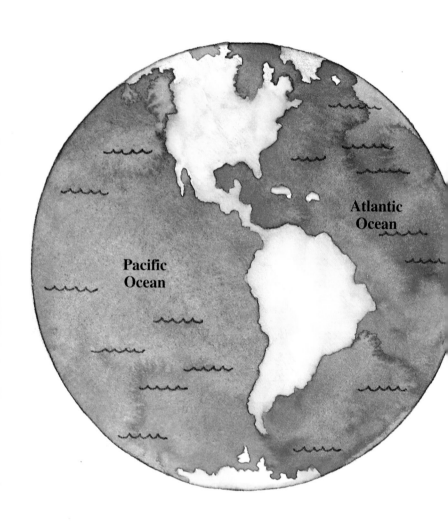

Atlantic
Ocean

Pacific
Ocean

Most water on the earth is in the oceans.

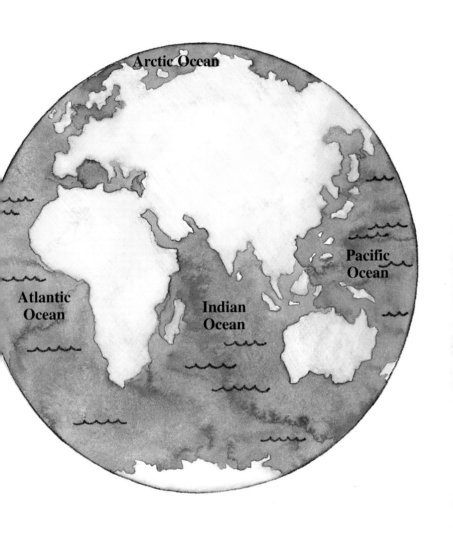

The oceans are salty.

The sun shines on the salty oceans.
It warms the water.

The warmth makes tiny bits of water
   called **molecules** (MOL-uh-kyools)
   rise out of the ocean.
The salty bits are left behind.
The water molecules form a **gas**.

The gas is like an invisible cloud.
It is called **water vapor**.
The water vapor disappears into the air.

You can't see the water vapor.
You can't smell it.
You can't feel it.
But water vapor is in the air.
When water becomes water vapor, we
   say it **evaporates** (e-VAP-uh-rates).

water vapor (invisible)

sun

heat

9

Water evaporates all around you.

Look for puddles after a rain.

The next day the puddles are gone.

Wash the chalkboard.
In a few minutes the board is dry.

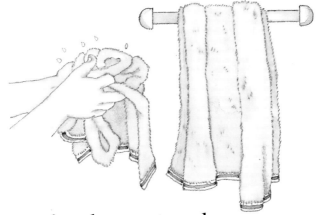

Dry your hands on a towel.
Look at the towel a few hours later.
The towel is not wet at all.

The water has evaporated.

Winds lift the water vapor.
They carry it high up in the air.

Here the air is cold.
The cold air chills the water vapor.
It changes into tiny drops of water.

When water vapor becomes water drops, we say it **condenses** (kon-DEN-ses).

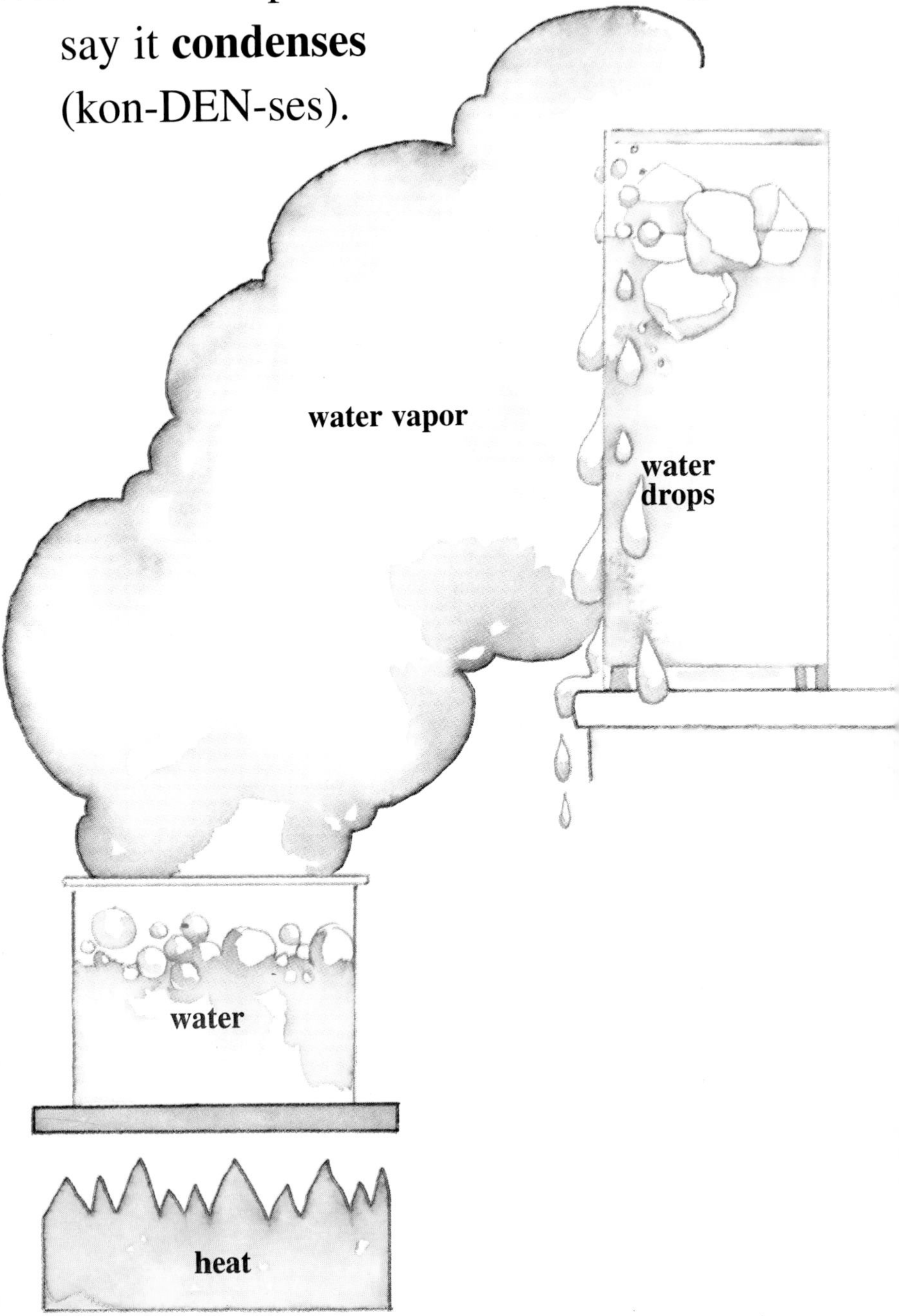

water vapor

water drops

water

heat

Water vapor condenses all around you.

Breathe out on a cold day.
The cold air chills your warm breath.
The water vapor in your breath turns into
a cloud of water drops.

water vapor (steam)

water vapor condenses on mirror

Take a hot bath.

The warm water causes water vapor, or **steam**, to form in the air.

The water vapor condenses on the cool bathroom mirror.

You fill a glass with milk.

The milk makes the glass cold.

The water vapor in the air condenses on the outside of the cold glass.

cloud

Up in the sky, billions of tiny water drops
come together.
They form a **cloud**.

The drops in the cloud bump into each
other.
They stick together.
The water drops grow bigger and bigger.

Sometimes the drops get too heavy to
stay up in the air.
Down they come as rain.
Rainwater is fresh water.
It is not salty like ocean water.

The rain makes the streets wet.
Puddles form.
Ponds and lakes and rivers fill with water.

Sometimes the air around a cloud is very, very cold.

The tiny water drops freeze into ice crystals.

The ice crystals stick together to form **snowflakes**.

The snowflakes fall as **snow**.

The snow may land on a mountain.
The snowflakes pile up.
They squeeze together.
The snow becomes solid ice.

The ice may slowly slide down the
   mountain and into a valley.
Slowly sliding ice is called a **glacier**
   (GLA-sher).

The glacier slides into an ocean.
The end of the glacier breaks off.
It becomes an **iceberg** (ISE-burg).
The iceberg floats into warmer waters.
And it melts.
Much of today's water is from glaciers
     that melted long, long ago.

Sometimes raindrops fall through very
    cold air.
The drops freeze as they fall.
They become little bits of ice.
Frozen raindrops are called **sleet**.

Sleet knocks against windows.
It clicks on the pavement.

Sometimes strong winds blow the sleet up again.

The winds sweep the sleet back into the clouds.

Up and down go the bits of ice.
The ice bumps into drops of water.
The water freezes.
The bits get bigger and bigger.
They grow into little balls of ice.

Finally the balls of ice are too heavy for the wind to carry.
They fall through the cold air.

The icy balls crash down to the earth.
The falling ice is called **hail**.
The lumps of ice are called **hailstones**.
Hailstones can be very big.
Some are as big as baseballs!

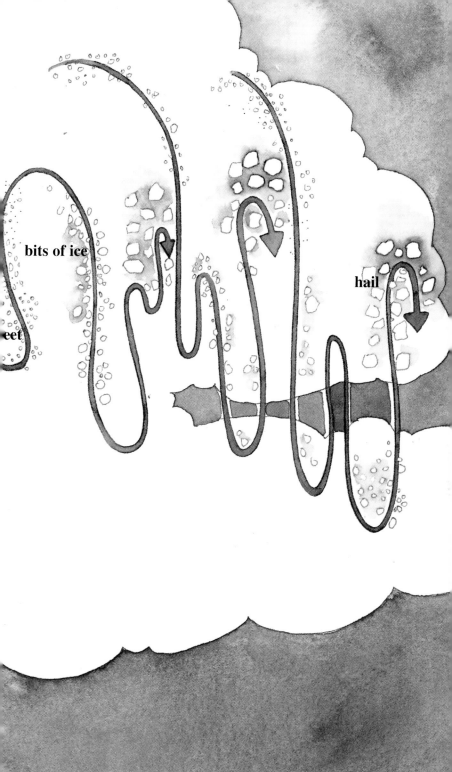

bits of ice

hail

eet

Sometimes the winds are not very strong.
The water vapor stays near the ground.
Maybe the ground is cool.
Then the vapor forms a low cloud.
We call such a cloud **fog**.

You can see the separate drops of water
    in fog.
Look for them on windowpanes.
Look for them on cars.
Fog makes everything look and feel wet.

Rain, snow, sleet, and hail.
They are all forms of water.
They

—fall from the sky

—mostly drop into oceans, lakes, and rivers

—evaporate from the earth

—condense in the air

—and fall from the sky all over again!

Round and round the water goes.
This is the water cycle.
It gives us the water we need.

water cycle

How does the water get to your home?

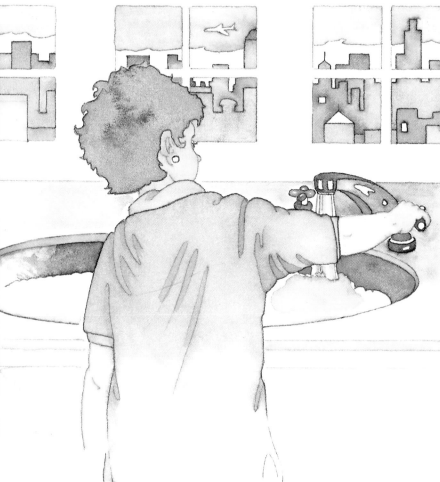

Suppose you live in a big city.
You turn on a faucet.
Out comes fresh, clean water.
Where does your water come from?

Most likely your water comes from a
 river or lake.
This water is then usually stored in a
 reservoir (REZ-er-vwar).
A reservoir is like a lake.

Reservoir water may not be clean.
In the water there may be
 —mud and dirt
 —fish, plants, and trash
 —chemicals and germs.

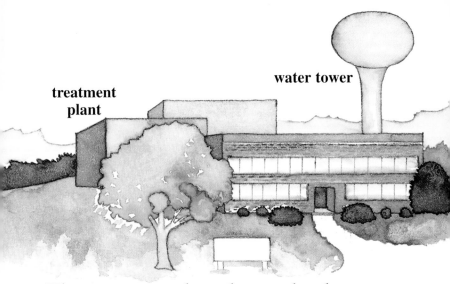

treatment plant

water tower

The water needs to be made clean
   enough to drink.
It goes to a treatment plant.

screen

Machines pump the water through a
   screen.
This takes out big pieces of trash.

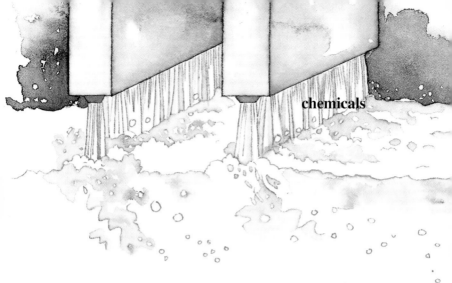

chemicals

Machines add chemicals to the water.
This makes smaller pieces settle to the
bottom.

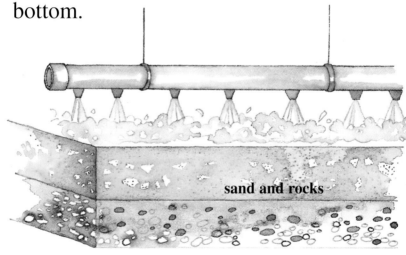

sand and rocks

Machines let the water drip through sand
and rocks.
This gets rid of any bits left in the water.

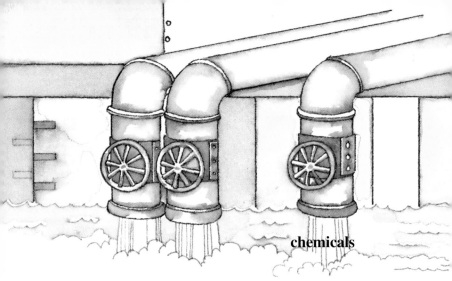

chemicals

Machines pour other chemicals into the water.
This kills germs in the water.

Machines spray the water into the air.
This gives the water a good smell and taste.

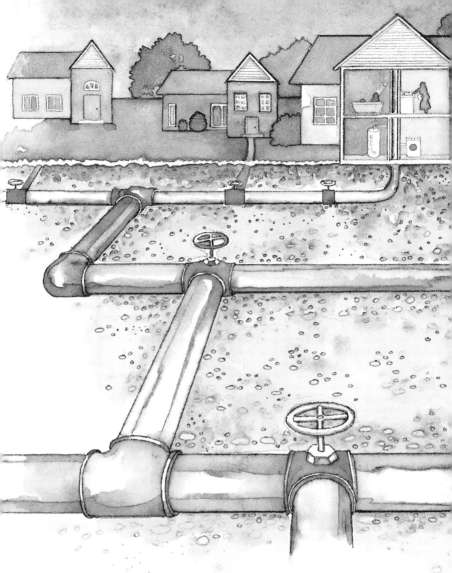

Next the water passes into big pipes.
From the big pipes, the water goes into
    small pipes.
The small pipes carry the water into your
    house.

Suppose you live in the country or in a
   small town.
Where does your water come from?

Most likely your water comes from a
   well.
A well is a deep hole in the ground.
It reaches water far below the ground.

Underground water comes from rain.
Some rainwater soaks through the soil.
It forms big pools of water under the
   ground.

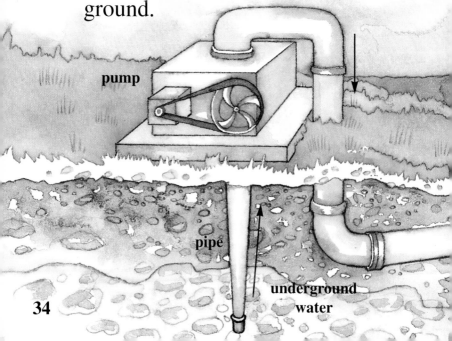

pump

pipe

underground
water

34

Diggers drill down into the ground to
make a well.
They push a pipe into the well.
Finally they reach the pool of water.
Then they put a pump at one end of the
pipe.
The pump pulls the water up through
the pipe.

Well water is usually very clean.
It may go right into your house.

water to house

Or the well water may be shared by a
   whole town.
Then the water first goes to a pumping
   station.
There it may be treated with chemicals.
It then travels through pipes to houses all
   over the town.
Some of the water may go to a water
   tower to be stored.

water tower

pumping
station

pump

chemicals

water flows
through pipes

underground
water

Your water may come from a well.
Or it may come from a reservoir.
Either way, the water is fresh and clean.

You use the water for
    —drinking
    —cleaning
    —cooking
    —bathing
    —and carrying away wastes.

Used water is no longer clean.
It is called **sewage** (SOO-ij).

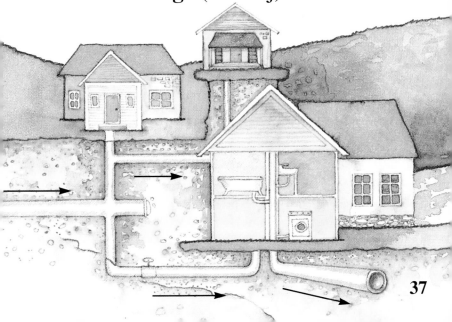

Many people live in cities.
Their sewage goes into pipes.
The pipes go to a sewage treatment plant.
At the plant
     —screens catch big pieces of solid
       matter
     —machines grind up leftover solids

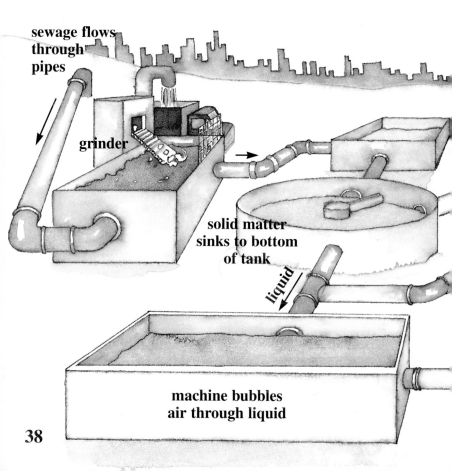

sewage flows
through
pipes

grinder

solid matter
sinks to bottom
of tank

liquid

machine bubbles
air through liquid

—the solid matter sinks to the bottom
—bacteria break down the solids
—machines bubble air through the
  liquid
—and chemicals break down any
  poisons in the liquid.

The cleaned water leaves the plant.
The water flows into a lake, river, or
  ocean.

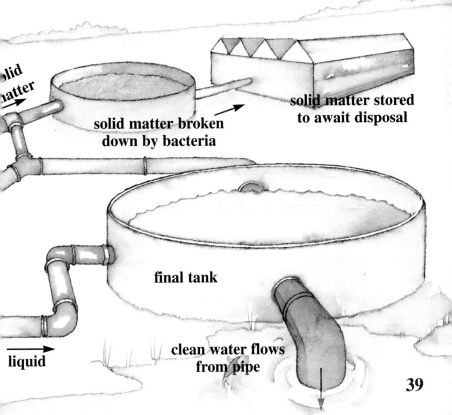

solid
matter

solid matter broken
down by bacteria

solid matter stored
to await disposal

final tank

liquid

clean water flows
from pipe

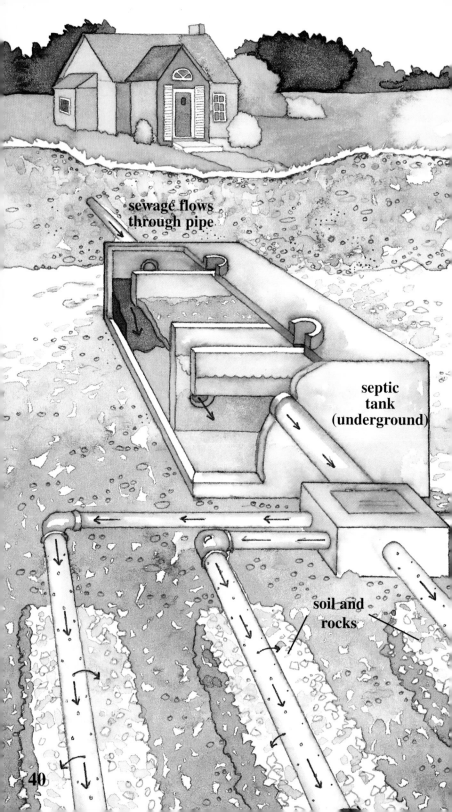

sewage flows
through pipe

septic
tank
(underground)

soil and
rocks

40

Other people live outside the cities.
Their sewage goes into **septic** (SEP-tik)
 tanks.
The tanks are under the ground.
Bacteria attack the solid matter.

The watery part trickles out.
It drips down through the soil and rocks.
As it drips it gets clean.
The cleaned water runs into an
 underground pool.
The pool also holds the dripped-down
 rainwater.

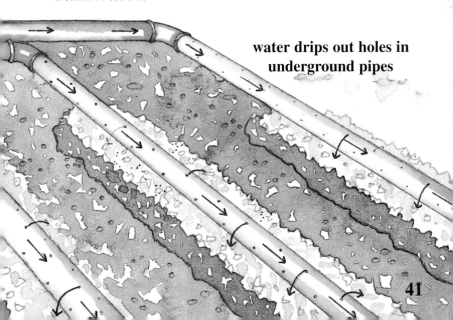

**water drips out holes in
underground pipes**

We all need lots of fresh, clean water.
Most people have enough for drinking,
    cleaning, and cooking.
Most farmers have enough to grow their
    crops.
Most factories have enough to make their
    products.

But more and more people are living on
the earth.

They need more and more water.

This causes problems:

   —too little clean water

   —too much dirt in our lakes, rivers,
   oceans, and reservoirs

   —too little money to clean all the
   dirty water.

How can you help save our water?

Tell an adult if you see sewage being
dumped into rivers, lakes, or oceans.
Ask a parent to check your home for
water leaks.

Help clean up beaches.

Shut off the water faucet while brushing
your teeth.

Take quick showers.

Ask an adult to put watersavers in toilets
and showerheads.

water cycle

Every drop of water is on a never-ending
    journey
        —from the oceans
        —up into the air
        —down to the earth
        —and back to the oceans again.

Let's use water wisely.
Then there will be plenty of fresh, clean
water for everyone.

Index